THE MAGNETIC UNKNOWN
HANDBOOK OF URANUS

Understanding Uncharted Fields and

Planetary Impact

MARCUS T. HOOKS

COPYRIGHT

TABLE OF CONTENTS

INTRODUCTION

The Magnetic Unknown Handbook of Uranus

In the vast expanse of our solar system, each planet is a world unto itself, shaped by unique forces, histories, and compositions that distinguish it from every other celestial body. Among the giants, Uranus stands out as one of the most enigmatic and least understood. As the seventh planet from the Sun, Uranus occupies a distant and chilly part of our cosmic neighborhood, veiled in shades of blue-green and hidden by layers of clouds. But beyond its striking hue lies a world full of peculiarities, from its extreme tilt and faint ring system to its complex, asymmetrical magnetic field that defies conventional understanding. In recent years, scientists have come to realize that much of what we thought we knew about Uranus might be incorrect—or, at the very least, incomplete. New interpretations of decades-old data, along with fresh insights from advanced simulations, are slowly peeling back the layers of mystery surrounding this icy giant.

The Magnetic Unknown Handbook of Uranus is a journey into one of the most intriguing aspects of this distant world: its magnetic field. Unlike Earth's relatively stable magnetic field, which aligns closely with its axis of rotation, Uranus's magnetosphere is both tilted and offset in ways that puzzle scientists. This "magnetic unknown" holds clues to the planet's internal structure, its atmospheric dynamics, and its interactions with the Sun and surrounding space. These magnetic forces create a chaotic environment, sparking phenomena unlike anything seen on other planets. Through this book, we'll delve into the latest theories, discoveries, and questions surrounding Uranus's magnetosphere and explore what they mean for our broader understanding of planetary science and the quest to uncover life-supporting conditions beyond Earth.

The Mystery Begins: Uranus as a Planetary Oddity

Since its discovery in 1781 by astronomer William Herschel, Uranus has fascinated scientists and astronomers alike. Initially mistaken for a star or a comet, Uranus was later recognized as the first planet to be discovered with the aid of a telescope. This discovery expanded our understanding of the solar system's boundaries and sparked a deeper interest in the outer planets. But as telescopes and space probes grew more advanced, Uranus proved to be more complex than initially thought. Unlike any other planet in the solar system, Uranus is tipped on its side, rotating almost perpendicular to the plane of its orbit. This extreme axial tilt of nearly 98 degrees means that Uranus doesn't experience seasons in the way that other planets do; instead, its poles take turns facing the Sun for long periods, creating a highly unusual seasonal cycle.

This unique orientation also has implications for Uranus's magnetic field. The planet's magnetic poles are significantly misaligned with its rotational axis, leading to a lopsided and constantly shifting magnetosphere. This chaotic magnetic field structure raises questions about Uranus's internal composition, the origins of its tilt, and the processes that drive its magnetosphere. Unlike the magnetic fields of Earth, Jupiter, or Saturn, Uranus's magnetic field appears to originate from a different layer within the planet, possibly from a highly conductive "ocean" of water, ammonia, and methane deep below the surface. This configuration, however, is far from fully understood, making Uranus's magnetic field a compelling area of study for scientists who seek to understand the diversity of planetary phenomena.

Voyager 2's Visit: A Glimpse into the Unknown

In 1986, NASA's Voyager 2 spacecraft became the first— and so far, only—probe to fly by Uranus, providing humanity with our first close-up view of this distant world.

During its brief encounter, Voyager 2 collected a wealth of data on Uranus, its moons, and its magnetic field. However, much of this data was difficult to interpret at the time, and with limited technology, many aspects of Uranus's magnetosphere remained elusive. Voyager 2's observations revealed an unexpected tilt in the planet's magnetic field, as well as a surprising degree of asymmetry. These findings suggested that Uranus's magnetic field was far more complex than previously imagined, sparking curiosity and opening the door to new lines of inquiry.

For decades, Voyager 2's data was the primary source of information on Uranus's magnetosphere. However, recent technological advances have allowed scientists to revisit this data with fresh perspectives. By reanalyzing Voyager's measurements and incorporating new computational models, researchers have begun to uncover surprising details that were overlooked or misinterpreted in the original analysis. For instance, scientists now suspect that Voyager 2's encounter with Uranus occurred during a rare solar wind event, which temporarily compressed the planet's magnetosphere. This compression may have skewed Voyager's data, leading to an incomplete understanding of the magnetic environment around Uranus. As a result, what we thought we knew about Uranus's magnetosphere could be substantially different from its true nature.

The Significance of Uranus's Magnetosphere

A planet's magnetic field is not merely a byproduct of its composition or rotation; it plays an essential role in shaping the planet's interactions with its environment, protecting it from harmful solar radiation, and influencing its atmospheric dynamics. For Earth, the magnetic field acts as a protective shield, deflecting solar wind and cosmic rays, which helps sustain life on the surface. Similarly, Uranus's magnetosphere shields it from the harsh conditions of space, but the way it functions is unlike anything seen in our solar system.

Uranus's magnetosphere is both tilted and offset from the planet's center, resulting in a field that constantly changes as the planet rotates. This dynamic, unstable magnetic environment gives rise to phenomena like powerful auroras and intense magnetic storms, although their exact nature remains poorly understood. Additionally, this irregular magnetosphere impacts Uranus's moons, particularly in terms of their potential to host subsurface oceans. Some of Uranus's moons, such as Titania, Oberon, and Miranda, are thought to contain layers of liquid water beneath their icy surfaces. The presence of a protective magnetic field could shield these moons from harmful radiation, creating stable conditions that might support microbial life.

For scientists, studying Uranus's magnetosphere is not just about understanding this one planet; it's about expanding our knowledge of planetary systems as a whole. The behavior of Uranus's magnetic field provides a natural laboratory for exploring the diversity of magnetic fields and their effects across the solar system. Insights gained from studying Uranus may even inform our understanding of exoplanets, which are likely to exhibit similarly diverse magnetic behaviors.

The Quest for Knowledge: Future Missions and Technological Advances

Despite its distance and relative isolation, Uranus has become a subject of increasing interest for future space missions. With advancements in technology, scientists are now better equipped to explore Uranus and its moons in ways that were previously impossible. A dedicated mission to Uranus could involve an orbiter equipped with advanced magnetometers, spectrometers, and imaging devices that could continuously study the planet's magnetosphere and atmospheric composition. Such a mission could revolutionize our understanding of Uranus's magnetic field, revealing details about its origin, its impact on the surrounding environment, and the interactions between the planet and its moons.

A new mission to Uranus would also allow scientists to study the planet over an extended period, observing how its magnetic field changes with seasonal shifts and solar activity. By tracking these changes, researchers could gain a deeper understanding of the magnetic forces at play and build a more complete picture of Uranus's complex magnetic environment. This, in turn, would contribute to a broader understanding of the diversity of planetary magnetic fields and how they evolve over time.

Additionally, studying Uranus's magnetosphere could yield insights into the potential habitability of its moons. With evidence suggesting that some of these moons might have subsurface oceans, the possibility of microbial life becomes a tantalizing prospect. If Uranus's magnetic field provides a stable, protective environment, it could create conditions favorable for life, albeit on a microscopic scale. Investigating these moons in greater detail would be a groundbreaking step in the search for life beyond Earth, expanding the scope of astrobiology to some of the most remote regions of our solar system.

What Lies Ahead in *The Magnetic Unknown Handbook of Uranus*

The Magnetic Unknown Handbook of Uranus is your guide to the latest discoveries, theories, and questions surrounding the magnetic complexities of this distant world. Each chapter will delve into different aspects of Uranus's magnetic field, from the initial discoveries made by Voyager 2 to the re-examination of data that has reshaped our understanding of the planet. We'll explore the role of Uranus's magnetosphere in shaping its atmospheric dynamics, the influence it has on its moons, and the broader implications for planetary science.

We'll also consider the role of magnetic fields in planetary evolution, how they affect atmospheres, and what they can reveal about a planet's interior structure. As we journey through the magnetic unknown of Uranus, you'll gain insights into the extraordinary diversity of magnetic phenomena in the solar system and how these phenomena shape the environments of planets and moons alike.

As you embark on this journey, keep in mind that Uranus, once considered one of the solar system's "quiet giants," is proving to be a dynamic world full of surprises. Its magnetic field, a source of both fascination and mystery, serves as a reminder of the complexities that lie beyond our everyday experiences on Earth. With each new discovery, we come one step closer to understanding not only Uranus but also the forces that govern our solar system and the broader universe.

Through *The Magnetic Unknown Handbook of Uranus*, we invite you to explore these mysteries with us. By examining the unique characteristics of Uranus's magnetosphere, we gain a fresh perspective on the diversity of planetary environments and the profound impact of magnetic forces. The journey into the magnetic unknown of Uranus is an invitation to push the boundaries of human knowledge, to question our assumptions, and to celebrate the spirit of discovery that drives us to reach beyond the known and into the vast, uncharted regions of space. Welcome to the exploration of Uranus and the magnetic mysteries that await us.

CHAPTER 1

Uranus Unveiled – A Historical Journey

The story of Uranus begins with its quiet existence as a faint point of light in the night sky, unnoticed by astronomers for centuries. Unlike the other planets visible to the naked eye, Uranus's dim appearance and slow movement across the heavens meant that ancient astronomers mistook it for a star. Its official discovery in 1781 by Sir William Herschel marked a turning point in astronomy, as Uranus became the first planet to be discovered using a telescope. This moment expanded humanity's understanding of the solar system's boundaries and spurred the search for other distant worlds.

1. Ancient Observations

Before Herschel's discovery, Uranus had been observed multiple times by astronomers who did not recognize it as a planet. The Greek astronomer Hipparchus and later Ptolemy cataloged it as a star. Even in the 17th century, when telescopic astronomy began, Uranus's planetary status remained undetected. This was largely due to its slow movement against the backdrop of stars, which made it indistinguishable from other celestial objects.

In hindsight, the planet was observed at least 20 times before Herschel's identification. One notable example was John Flamsteed, the first Astronomer Royal of England, who documented Uranus in 1690 and cataloged it as a star, "34 Tauri." His observations contributed to the eventual recognition of Uranus as a planet.

2. Herschel's Discovery

On March 13, 1781, Sir William Herschel was conducting a telescopic survey of the night sky when he noticed an unusual object. Initially believing it to be a comet, Herschel noted its peculiar shape and motion, which differed from stars. Continued observations confirmed that the object followed a planetary orbit, and Herschel eventually presented his findings to the Royal Society. Although Herschel proposed naming the planet "Georgium Sidus" (George's Star) in honor of King George III, the name Uranus, derived from the Greek god of the sky, was later adopted to align with the classical naming tradition of planets.

The discovery of Uranus was a monumental achievement, extending the known boundaries of the solar system and setting the stage for further explorations. It underscored the importance of telescopes in advancing our understanding of the universe and marked the beginning of modern planetary science.

3. Advancements in Uranus Observation

Throughout the 19th and early 20th centuries, advances in telescopic technology allowed astronomers to learn more about Uranus. Its unusual axial tilt of approximately 98 degrees—causing the planet to rotate almost on its side—was identified, along with its faint ring system and numerous moons. Observations of its orbit also revealed irregularities, suggesting the gravitational influence of an unknown object beyond Uranus. This led to the discovery of Neptune in 1846, further highlighting the interconnectedness of celestial mechanics within the solar system.

By the mid-20th century, Uranus was recognized as one of the solar system's ice giants, distinguished by its composition of hydrogen, helium, water, ammonia, and methane. However, much about the planet, including its atmospheric dynamics, magnetic field, and internal structure, remained elusive, awaiting the arrival of spacecraft capable of providing detailed data.

The Voyager 2 Mission: A Detailed Account of Discoveries

The true unveiling of Uranus began with NASA's Voyager 2 spacecraft, which remains the only probe to visit the planet. Launched in 1977, Voyager 2 was part of a twin-mission project designed to explore the outer planets. While its primary targets were Jupiter and Saturn, its trajectory allowed it to continue to Uranus and eventually Neptune, providing humanity with its first close-up views of these distant worlds.

1. The Uranus Flyby

On January 24, 1986, Voyager 2 flew within 50,600 miles (81,500 kilometers) of Uranus's cloud tops, capturing unprecedented images and data. This brief encounter, lasting only a few hours, provided a wealth of information about Uranus's atmosphere, moons, rings, and magnetic field. Despite its short duration, the flyby transformed our understanding of Uranus and its place in the solar system.

2. Discoveries About Uranus

Voyager 2's observations revealed several key aspects of Uranus, many of which challenged prior assumptions:

- **Atmosphere**: Voyager 2 detected an atmosphere dominated by hydrogen and helium, with significant amounts of methane that gave the planet its blue-green color. The probe also identified the coldest temperatures in the solar system, with Uranus's upper atmosphere reaching as low as -371°F (-224°C). The absence of significant heat from the planet's core remains a mystery, as other gas giants like Jupiter and Saturn emit substantial amounts of internal heat.

- **Magnetic Field**: One of the most striking discoveries was Uranus's magnetic field, which is both tilted at a 59-degree angle relative to its rotational axis and offset from the planet's center. This asymmetrical and dynamic magnetosphere challenged conventional models of planetary magnetism and hinted at an unconventional internal structure.

- **Rings and Moons**: Voyager 2 confirmed the presence of 11 previously known rings and discovered two new ones, shedding light on the structure and composition of Uranus's ring system. It also observed 10 new moons, bringing the total known to 27. Among these moons, Miranda stood out for its dramatic surface features, including massive cliffs and varied terrains, suggesting a history of geological activity.

- **Rotation and Seasons**: The spacecraft documented Uranus's unique rotation and extreme seasonal variations, caused by its axial tilt. These findings raised questions about how such a tilt could have occurred, with theories suggesting a massive collision early in the planet's history.

3. The Legacy of Voyager 2

Voyager 2's flyby of Uranus marked a turning point in planetary science, offering the first detailed glimpse of a world previously seen only as a distant, blurry dot. Its discoveries reshaped our understanding of Uranus and provided a foundation for future research. However, the mission also highlighted the need for more comprehensive exploration, as many questions about Uranus and its system remained unanswered due to the limitations of a single, short-duration flyby.

Mysteries and Gaps: Why Uranus Remains One of the Least Understood Planets

Despite Voyager 2's groundbreaking discoveries, Uranus remains one of the most mysterious planets in our solar system. Its distance from Earth, unique characteristics, and the lack of subsequent missions have left many gaps in our knowledge.

1. Limited Data from Voyager 2

While Voyager 2 provided invaluable insights, its brief encounter with Uranus allowed only a snapshot of the planet's complex system. Key areas, such as the detailed composition of its atmosphere, the dynamics of its magnetosphere, and the geological processes shaping its moons, remain poorly understood. Moreover, Voyager 2's instruments, though advanced for their time, were not designed to address all the questions scientists now have about Uranus.

2. Challenges of Observing Uranus from Earth

Uranus's distance from the Sun (about 1.8 billion miles) and its relatively featureless appearance make it challenging to study from Earth. Ground-based telescopes and even space-based observatories like the Hubble Space Telescope struggle to capture detailed images or analyze its atmosphere. Unlike Jupiter or Saturn, which have dramatic cloud patterns and storms, Uranus's subdued features make it harder to track atmospheric changes and dynamics.

3. Unresolved Questions About Its Magnetic Field

Uranus's magnetosphere remains one of its greatest enigmas. The tilt and offset of its magnetic field suggest a highly unusual internal structure, possibly involving conductive layers of water, ammonia, and methane rather than a conventional metallic core. However, without additional data, scientists cannot confirm these hypotheses or fully understand the processes driving Uranus's magnetic behavior.

4. The Need for a Dedicated Mission

One of the main reasons Uranus remains poorly understood is the absence of follow-up missions. Unlike Mars, Jupiter, and Saturn, which have been visited by multiple spacecraft, Uranus has only been studied up close by Voyager 2. A dedicated mission to Uranus, equipped with modern instruments, could provide answers to many of the lingering questions about its atmosphere, magnetosphere, rings, and moons. Proposed missions, such as orbiters or atmospheric probes, have yet to materialize due to budgetary and logistical challenges, leaving Uranus largely unexplored.

The exploration of Uranus is a testament to humanity's curiosity and ingenuity, revealing a planet as mysterious as it is unique. From its discovery in 1781 to the Voyager 2 flyby in 1986, Uranus has challenged our understanding of planetary science and offered tantalizing glimpses into the complexities of the outer solar system. Yet, much remains unknown. Its tilted axis, frigid atmosphere, dynamic magnetosphere, and intriguing moons continue to spark questions that only future missions can answer.

CHAPTER 2

The Magnetic Puzzle – New Findings on Uranus's Magnetosphere

When NASA's Voyager 2 spacecraft flew past Uranus in 1986, it was humanity's first and only close encounter with the seventh planet from the Sun. The data collected during the brief flyby provided fascinating insights into Uranus's magnetic field, revealing a tilted, asymmetric magnetosphere that was unlike anything observed in the solar system. However, decades after the Voyager 2 mission, scientists have begun to question whether the snapshot captured during the flyby accurately represents Uranus's typical magnetic environment.

Recent studies have revisited the Voyager 2 data with advanced computational models and a deeper understanding of solar wind interactions. Researchers now believe that Voyager 2's measurements were taken during a rare and intense solar wind event—a high-speed stream of charged particles emitted by the Sun. This event likely compressed Uranus's magnetosphere to a fraction of its usual size, distorting the measurements and creating the illusion of a smaller, less complex magnetic field.

1. Understanding the Solar Wind's Role

The solar wind is a continuous stream of charged particles, primarily electrons and protons, that flows outward from the Sun. This plasma interacts with the magnetic fields of planets, shaping their magnetospheres. On Earth, the solar wind creates phenomena like auroras by colliding with atmospheric particles, while the magnetosphere deflects most of these high-energy particles, protecting the planet from harmful radiation.

For Uranus, the interaction between its magnetosphere and the solar wind is particularly complex due to the planet's extreme axial tilt and unique magnetic field orientation. Uranus rotates nearly on its side, with its magnetic field tilted at a 59-degree angle to its rotational axis. This configuration creates a highly dynamic magnetosphere that shifts dramatically as the planet rotates. The impact of the solar wind on such a system is not straightforward, making it challenging to interpret Voyager 2's data.

2. The Rare Solar Wind Event During Voyager 2's Flyby

During Voyager 2's closest approach, scientists now believe that Uranus's magnetosphere was compressed to about 20% of its typical size by a rare, high-pressure solar wind event. Such events occur when streams of high-speed particles from the Sun collide with slower-moving regions of the solar wind, creating a shockwave that propagates through the solar system. For Uranus, this event temporarily reduced the extent of its magnetosphere, exposing regions that would normally be shielded and altering the dynamics of its magnetic field.

This realization has led researchers to question many of the conclusions drawn from Voyager 2's data. For example, the apparent lack of plasma and energetic particles in Uranus's magnetosphere may have been a temporary effect of the solar wind event, rather than a permanent feature. Similarly, the intense belts of energetic electrons observed by Voyager 2 could have been artificially enhanced by the compression of the magnetosphere, rather than representing a typical state.

3. Advancing Our Understanding Through Reanalysis

By revisiting the Voyager 2 data with modern tools and techniques, scientists are beginning to correct these distortions and develop a more accurate picture of Uranus's magnetic field. Computational models that simulate the interactions between the solar wind and Uranus's magnetosphere have been instrumental in this process, revealing how the field behaves under different conditions.

These models suggest that Uranus's magnetosphere is far more complex and dynamic than previously thought, with regions of intense activity that shift and evolve over time.

Revised Understanding of Uranus's Magnetic Field: Comparisons with Jupiter and Saturn

Magnetic fields are fundamental features of planetary systems, providing insights into a planet's interior composition, atmospheric dynamics, and interactions with its environment. While magnetic fields are common among planets in the solar system, Uranus's magnetic field is uniquely peculiar, differing significantly from those of Jupiter, Saturn, and Earth.

1. Earth's Stable Magnetosphere vs. Uranus's Dynamic Field

Earth's magnetic field is generated by the motion of molten iron and nickel in its outer core, creating a relatively stable dipole field that aligns closely with the planet's rotational axis. This alignment results in a consistent, protective magnetosphere that shields the planet from solar wind and cosmic radiation.

In contrast, Uranus's magnetic field is both tilted and offset. Its magnetic poles are misaligned with its rotational axis by

59 degrees, and the field itself is offset from the planet's center by about one-third of its radius. This asymmetry creates a highly irregular magnetosphere that changes dramatically as the planet rotates. Unlike Earth, where the magnetosphere forms a relatively uniform shape, Uranus's magnetosphere is lopsided and fluctuates between open and closed configurations, exposing different regions of the planet to solar wind.

2. Jupiter and Saturn: Giant Magnetic Fields

Jupiter and Saturn, the gas giants closest to Uranus in the solar system, have magnetic fields that are far more powerful and symmetrical. Jupiter's magnetic field, for example, is the strongest among the planets, generated by metallic hydrogen in its interior. This powerful field creates an enormous magnetosphere that extends millions of miles into space, protecting Jupiter and its moons from solar wind.

Saturn's magnetic field, while weaker than Jupiter's, is still remarkably symmetrical and closely aligned with its rotational axis. This alignment results in a well-defined magnetosphere that interacts predictably with the solar wind and its ring system.

Uranus, by comparison, is an outlier. Its tilted and offset magnetic field suggests a fundamentally different internal structure. Scientists hypothesize that Uranus's magnetic field is generated by a layer of conductive materials, such as water, ammonia, and methane, rather than a metallic core. These materials, under extreme pressure and temperature, may form a "superionic" state that conducts electricity and generates the magnetic field. However, this theory remains unconfirmed, as direct measurements of Uranus's interior are still unavailable.

3. Implications for Planetary Science

The peculiarities of Uranus's magnetic field challenge traditional models of planetary magnetism. Understanding how Uranus's field is generated and maintained could provide new insights into the diversity of planetary systems and the processes that shape them. For example, studying Uranus could help scientists refine their understanding of magnetic fields on exoplanets, which may exhibit similar asymmetries and tilts.

Implications for Uranus and Beyond: Understanding the Magnetosphere's Role in the Planet's Environment

Uranus's magnetosphere is not just a scientific curiosity; it plays a vital role in shaping the planet's environment and influencing its moons. By studying Uranus's magnetic field, scientists can gain insights into the interactions between the planet, its atmosphere, and its surrounding space.

1. The Magnetosphere's Protective Shield

Like Earth, Uranus's magnetosphere acts as a shield, protecting the planet from the constant bombardment of charged particles in the solar wind. However, the irregular shape and dynamics of Uranus's magnetosphere mean that this protection is not uniform. Some regions are more exposed to solar wind than others, creating localized effects that may influence atmospheric composition and weather patterns.

The interaction between the solar wind and Uranus's magnetosphere also generates auroras, similar to the northern and southern lights on Earth. These auroras, however, are less well understood due to the limited observations available. Studying these phenomena could provide valuable insights into the dynamics of Uranus's magnetic field and its interactions with the solar wind.

2. Effects on Uranus's Moons

Uranus's magnetosphere extends far enough to encompass many of its moons, creating a dynamic environment that influences their surfaces and interiors. For moons like

Titania, Oberon, and Miranda, the magnetosphere may act as a protective barrier, shielding them from harmful solar radiation. This protection could create stable conditions that support subsurface oceans, raising the possibility of habitability.

Recent studies suggest that some of Uranus's moons may harbor liquid water beneath their icy crusts. The presence of a magnetosphere could help sustain these oceans by reducing radiation-induced ice loss and maintaining stable conditions. Understanding the interactions between Uranus's magnetosphere and its moons is therefore critical for assessing their potential to host life.

3. Broader Implications for Exoplanets and Space Exploration

The study of Uranus's magnetosphere has implications that extend beyond the solar system. Many exoplanets—planets orbiting stars outside our solar system—are thought to exhibit magnetic fields, but their diversity and dynamics remain poorly understood. Uranus serves as a natural laboratory for studying how unusual magnetic fields operate, offering clues that could help scientists interpret observations of exoplanets with similar characteristics.

Furthermore, the exploration of Uranus's magnetosphere underscores the need for dedicated missions to study the ice giants in greater detail. A new mission to Uranus, equipped with advanced instruments, could provide the data needed to confirm theories about its magnetic field, interior structure, and interactions with the solar wind. Such a mission would not only enhance our understanding of Uranus but also contribute to the broader goal of exploring the outer solar system and its potential for life.

Uranus's magnetosphere is a puzzle that continues to challenge and intrigue scientists. From the distorted measurements caused by a rare solar wind event during Voyager 2's flyby to the ongoing efforts to understand its tilted and offset magnetic field, Uranus remains a planet of magnetic mysteries. By revisiting Voyager 2's data and incorporating new technologies, researchers are beginning to unravel the complexities of this unique magnetosphere, offering fresh insights into the forces that shape Uranus and its environment.

As we deepen our understanding of Uranus's magnetic field, we uncover not only the secrets of a distant planet but also the broader dynamics of planetary magnetism and its role in shaping worlds throughout the universe. The study of

Uranus's magnetosphere is a testament to the enduring power of curiosity and exploration, reminding us that even the coldest, most distant corners of the solar system have stories to tell.

CHAPTER 3

Reimagining Uranus's Moons – An Ocean Beneath the Ice?

Uranus, the seventh planet from the Sun, is not only enigmatic due to its extreme axial tilt and unique magnetic field but also because of its family of 27 known moons. These moons, composed largely of ice and rock, vary significantly in size, surface features, and potential for geological activity. Among them, Titania, Oberon, and Miranda stand out as particularly intriguing. Their distinct characteristics make them prime candidates for exploring the possibility of subsurface oceans and the potential for habitability within Uranus's system.

1. Titania: The Largest and Most Intriguing

Titania, the largest of Uranus's moons, measures approximately 1,578 kilometers (980 miles) in diameter, making it the eighth-largest moon in the solar system. Discovered by William Herschel in 1787, Titania is composed of roughly equal parts ice and rock, with a heavily cratered surface that suggests a long and complex geological history.

One of Titania's most fascinating features is its network of canyons and fault lines, which indicate past tectonic activity. These structures suggest that Titania experienced internal heating at some point in its history, possibly due to tidal forces exerted by Uranus. This heating could have created conditions favorable for maintaining liquid water beneath its icy crust, making Titania a prime candidate for harboring a subsurface ocean.

2. Oberon: A Moon of Contrasts

Oberon, the second-largest moon of Uranus, is slightly smaller than Titania, with a diameter of approximately 1,523 kilometers (946 miles). Like Titania, Oberon was discovered by William Herschel in 1787 and shares a similar composition of ice and rock. However, Oberon's surface is markedly different, with a higher density of craters and less evidence of tectonic activity. This suggests that Oberon has been geologically inactive for much of its history.

Despite its apparent inactivity, Oberon's composition and proximity to Uranus raise the possibility of a subsurface ocean. If tidal forces or residual heat from radioactive decay have persisted beneath its crust, these could create the necessary conditions for liquid water to exist. Further exploration is needed to determine whether Oberon's interior harbors hidden secrets.

3. Miranda: The Smallest and Most Active

Miranda, the smallest of Uranus's major moons, is also the most geologically diverse and intriguing. With a diameter of just 472 kilometers (293 miles), Miranda is known for its striking surface features, including massive cliffs, chevron-shaped ridges, and regions of varying terrain. These features suggest a history of intense geological activity, likely driven by tidal interactions with Uranus and other moons.

One of Miranda's most remarkable landmarks is Verona Rupes, the tallest known cliff in the solar system, which rises approximately 20 kilometers (12 miles) above the surface. This dramatic feature, along with Miranda's other unique structures, hints at a dynamic interior that may include a subsurface ocean. Recent studies suggest that Miranda's small size does not preclude the possibility of liquid water beneath its ice, particularly if tidal heating remains active.

The Magnetosphere's Protective Role: Supporting Conditions for Subsurface Oceans

A key factor in assessing the potential for oceans beneath the ice of Uranus's moons is the influence of the planet's magnetosphere. Uranus's magnetic field, while irregular and offset, extends far enough to encompass many of its moons, creating a dynamic environment that affects their surfaces and interiors. Understanding how this magnetosphere interacts with the moons is critical to evaluating their habitability and the likelihood of liquid water beneath their crusts.

1. Shielding from Solar Radiation

Uranus's magnetosphere acts as a protective shield, deflecting charged particles from the solar wind and cosmic radiation. For moons like Titania, Oberon, and Miranda, this shielding reduces the rate of ice erosion and preserves the structural integrity of their surfaces. More importantly, it minimizes the radiation exposure to potential subsurface oceans, which could help maintain their stability over geological timescales.

The degree of protection provided by the magnetosphere varies depending on the moon's location and the dynamics of the magnetic field. Moons closer to Uranus, such as Miranda, are more likely to be shielded, while those farther away, like Titania and Oberon, experience greater exposure to solar wind. Understanding these variations is essential for assessing the habitability of these moons.

2. Induced Magnetic Fields

Subsurface oceans containing salts or other conductive materials can generate secondary, or induced, magnetic fields when exposed to Uranus's magnetosphere. These fields are detectable through spacecraft observations and can serve as direct evidence of liquid water beneath a moon's crust. For example, Jupiter's moon Europa and Saturn's moon Enceladus have shown signs of induced magnetic fields, confirming the presence of subsurface oceans.

Although Uranus's moons have not yet been studied in sufficient detail to detect such fields, their potential existence is a tantalizing possibility. Future missions equipped with sensitive magnetometers could investigate whether Titania, Oberon, or Miranda exhibit similar magnetic signatures, providing definitive proof of hidden oceans.

3. Energy Sources and Heating

The magnetosphere also influences the energy balance of Uranus's moons, particularly through interactions with charged particles. These interactions generate low levels of heat, which, combined with tidal forces and radioactive decay, could contribute to maintaining liquid water beneath the icy crusts. While this heating is less intense than that experienced by moons like Io or Europa, it may be sufficient to sustain subsurface oceans in Uranus's system.

Potential for Subsurface Oceans: Evidence and Implications

The possibility of subsurface oceans on Uranus's moons is a subject of growing interest, fueled by discoveries of similar oceans on other icy worlds in the solar system. While direct evidence is lacking, several lines of research suggest that Titania, Oberon, and Miranda could harbor liquid water beneath their icy exteriors.

1. Geological Evidence

The surface features of Uranus's moons provide important clues about their internal structures. Titania's canyons and Miranda's chaotic terrain, for example, suggest past or ongoing geological activity that may be linked to subsurface oceans. These features are consistent with processes such as cry volcanism or tectonics, which require internal heat and, in some cases, liquid water.

For Oberon, the evidence is less direct, but its composition and size suggest that it could have experienced enough internal heating to sustain a liquid layer in the past. If residual heat persists, this ocean could still exist today.

2. Tidal Heating

Tidal forces exerted by Uranus and neighboring moons are another potential source of heat for maintaining subsurface oceans. As these moons orbit Uranus, gravitational interactions cause their interiors to flex and generate frictional heat. This process, known as tidal heating, is a well-established mechanism for sustaining oceans on icy moons like Europa and Enceladus.

For Uranus's moons, the intensity of tidal heating depends on their orbital configurations and distances from the planet. Miranda, being the closest of the three, is most likely to experience significant tidal heating, while Titania and Oberon may benefit from weaker but still meaningful effects.

3. Implications for Habitability

The presence of subsurface oceans on Uranus's moons would have profound implications for the search for life beyond Earth. Liquid water is a key ingredient for life as we know it, and the discovery of such environments on Titania, Oberon, or Miranda would expand the range of potentially habitable worlds in the solar system.

Subsurface oceans are particularly intriguing because they are shielded from the harsh conditions of space, including extreme temperatures and radiation. If these oceans contain the necessary chemical ingredients, such as organic compounds and energy sources, they could provide suitable conditions for microbial life to exist.

4. Future Missions to Confirm Oceans

To confirm the existence of subsurface oceans on Uranus's moons, future missions will need to employ advanced technologies capable of probing beneath their icy crusts. Instruments such as ice-penetrating radar, magnetometers, and thermal imagers could provide direct evidence of liquid water. Additionally, landers or orbiters could analyze surface compositions and search for signs of cryovolcanic activity, which would indicate the presence of subsurface reservoirs.

NASA and other space agencies have proposed missions to Uranus as part of their long-term exploration strategies. These missions could revolutionize our understanding of the planet's moons and their potential for hosting life, making them a priority for future exploration.

Uranus's moons, particularly Titania, Oberon, and Miranda, represent some of the most intriguing and underexplored worlds in the solar system. Their unique geological features, combined with the potential influence of Uranus's magnetosphere, suggest that they may harbor subsurface oceans beneath their icy exteriors. These hidden oceans, if confirmed, could redefine our understanding of habitability and the potential for life beyond Earth.

While much remains unknown, the evidence gathered so far highlights the need for dedicated missions to Uranus and its moons. By probing these icy worlds with advanced instruments, scientists can uncover the secrets of their interiors, assess their habitability, and expand humanity's knowledge of the solar system's outer reaches. The search for oceans beneath the ice of Uranus's moons is not just a quest for scientific discovery—it is a journey into the unknown, where the possibility of finding life in the most unexpected places continues to inspire and captivate.

CHAPTER 4

Miranda – The Hidden Gem with Potential for Life

Miranda, the smallest and innermost of Uranus's five major moons, is a world of dramatic contrasts and mysteries. Despite its diminutive size—measuring only 472 kilometers (293 miles) in diameter—Miranda's surface features some of the most extraordinary landscapes in the solar system. The moon's terrain is a patchwork of massive cliffs, smooth plains, ridged belts, and chaotic regions, each telling a story of a dynamic geological past. While much of Uranus's system remains shrouded in mystery due to limited exploration, Miranda stands out as a prime candidate for further study, particularly in the context of subsurface oceans and the potential for life. This chapter delves into Miranda's unique features, the evidence for a subsurface ocean beneath its icy crust, and the conditions that might make this moon hospitable to life.

Miranda's Unique Features: A Geologically Complex Landscape

Discovered in 1948 by Gerard Kuiper, Miranda has captured the fascination of scientists due to its starkly contrasting surface features. Voyager 2's 1986 flyby provided the only close-up images of the moon, revealing a landscape so diverse and unexpected that it has been described as a "geological Frankenstein." Unlike the smooth, monotonous terrains of some moons, Miranda's surface tells a story of upheaval, transformation, and possibly ongoing activity.

1. The Patchwork Terrain

Miranda's surface is divided into two primary types of terrain: ancient, heavily cratered regions and younger, less cratered areas with distinctive ridges and grooves. These contrasting terrains suggest a complex geological history marked by episodes of resurfacing and renewal. The older regions, heavily pockmarked with craters, hint at a long period of relative inactivity, while the smoother, younger areas indicate more recent geological processes.

Some of the most intriguing regions on Miranda are the coronae—oval-shaped formations characterized by concentric ridges and valleys. Three prominent coronae, named Arden, Elsinore, and Inverness, dominate Miranda's landscape, each spanning hundreds of kilometers. These features are thought to result from tectonic activity, possibly driven by internal heating or tidal forces.

2. Verona Rupes: The Tallest Cliff in the Solar System

Among Miranda's most famous landmarks is Verona Rupes, a towering cliff that rises approximately 20 kilometers (12 miles) above the surface. To put this into perspective, Verona Rupes is nearly three times the height of Mount Everest. This sheer drop is the result of tectonic activity and remains one of the most striking features discovered in the outer solar system. The existence of such a massive geological structure highlights the intense forces that have shaped Miranda's surface.

3. Evidence of Cry volcanism

Voyager 2's observations suggest that Miranda may have experienced cry volcanism—volcanic activity involving water or other volatile substances rather than molten rock. The presence of smooth plains and ridged belts indicates that material from Miranda's interior may have erupted onto the surface, resurfacing certain areas. This process could be linked to internal heating, which might also support the existence of a subsurface ocean.

4. A History of Disruption

Miranda's unique surface features have led scientists to hypothesize that the moon may have been shattered and reassembled multiple times during its history. This theory is supported by the chaotic arrangement of its terrains, which resemble a jigsaw puzzle of mismatched pieces. These disruptions could have been caused by impacts or gravitational interactions with Uranus and neighboring moons.

Evidence of a Subsurface Ocean: Unveiling Miranda's Hidden Depths

While Miranda's surface is fascinating, its interior may hold even greater secrets. Recent studies have raised the possibility that a liquid ocean exists beneath Miranda's icy crust, much like the oceans believed to exist on Europa, Enceladus, and other icy moons in the solar system.

1. Tidal Heating as a Driving Force

One of the primary mechanisms that could sustain a subsurface ocean on Miranda is tidal heating. As Miranda orbits Uranus, it is subjected to gravitational forces that cause its interior to flex and generate heat. This process, known as tidal heating, is a well-established phenomenon that also explains the subsurface oceans on moons like Europa and Enceladus.

For Miranda, tidal heating would be particularly intense due to its proximity to Uranus and its likely interactions with neighboring moons. The heat generated by this process could be sufficient to maintain a layer of liquid water beneath the moon's icy crust, even in the frigid temperatures of the outer solar system.

2. Conductivity and Induced Magnetic Fields

If Miranda's subsurface ocean contains salts or other conductive materials, it could generate an induced magnetic field in response to Uranus's magnetosphere. These induced fields have been detected on other ocean-bearing moons, providing direct evidence of liquid water. While no spacecraft has yet measured Miranda's magnetic environment in sufficient detail, future missions equipped with magnetometers could confirm the presence of such a field, offering definitive proof of a subsurface ocean.

3. Geological Clues from Surface Features

Miranda's ridged terrains and coronae provide indirect evidence of subsurface activity. These features suggest that material from the moon's interior has pushed upward, deforming the surface. Such processes are often associated with liquid reservoirs beneath the crust. If a subsurface ocean exists, it could be responsible for the resurfacing events that created Miranda's smoother regions and coronae.

4. Potential for an Insulating Ice Shell

The presence of an insulating ice shell over Miranda's subsurface ocean could help retain heat and maintain liquid water. This shell, composed of water ice and possibly other volatiles, would act as a barrier, preventing the ocean from freezing completely. The thickness of this shell would depend on the balance between heat generated by tidal forces and the rate of heat loss to space.

Conditions for Life: Could Miranda Support Habitable Environments?

The possibility of a subsurface ocean on Miranda raises the tantalizing question of whether this tiny moon could support life. While Miranda's environment is vastly different from Earth's, certain conditions could make it hospitable to microbial life or other simple organisms.

1. The Role of Liquid Water

Liquid water is one of the essential ingredients for life as we know it. If Miranda harbors a subsurface ocean, it provides a stable, long-term environment where life could potentially exist. Subsurface oceans are particularly promising because they are shielded from the harsh conditions of space, including extreme temperatures and radiation.

2. Energy Sources

In addition to liquid water, life requires a source of energy. For Miranda, tidal heating could serve as a primary energy source, driving chemical reactions within the subsurface ocean. These reactions could provide the energy necessary to sustain microbial ecosystems, similar to hydrothermal vents on Earth's ocean floor.

Hydrothermal activity is particularly important in the context of habitability. If Miranda's subsurface ocean is in contact with a rocky core, chemical interactions between water and rock could produce nutrients like hydrogen, methane, and sulphides'. These compounds could serve as the building blocks for life or sustain existing microbial communities.

3. Protective Magnetic Fields

Uranus's magnetosphere plays a critical role in shielding its moons from solar wind and cosmic radiation. For Miranda, this protective environment could help maintain the stability of its subsurface ocean and prevent radiation-induced damage to potential life forms. While Miranda's proximity to Uranus subjects it to intense gravitational forces, the magnetosphere provides a degree of protection that is crucial for habitability.

4. Comparisons with Other Ocean Worlds

Miranda's potential for habitability can be better understood by comparing it to other ocean worlds in the solar system. Europa and Enceladus, moons of Jupiter and Saturn respectively, are considered prime candidates for life due to their subsurface oceans and evidence of hydrothermal activity. While Miranda is smaller and less studied than these moons, the mechanisms that sustain its ocean could be similar.

5. Challenges and Limitations

Despite its potential, Miranda faces several challenges as a habitable environment. The moon's small size limits its ability to retain heat over long timescales, which could lead to the eventual freezing of its subsurface ocean. Additionally, the lack of direct evidence for hydrothermal vents or other energy sources makes it difficult to assess the moon's true habitability. Further exploration is needed to determine whether Miranda's subsurface ocean contains the necessary conditions for life.

Miranda, with its dramatic landscapes and hidden depths, represents one of the most intriguing worlds in the solar system. Its unique surface features, from towering cliffs to enigmatic coronae, hint at a dynamic past shaped by powerful geological forces. Beneath its icy crust, the possibility of a subsurface ocean raises exciting questions about the moon's potential for habitability and the conditions necessary to sustain life.

While much about Miranda remains speculative, the evidence gathered so far highlights its importance as a target for future exploration. A dedicated mission to Uranus and its moons, equipped with advanced instruments, could provide definitive answers about Miranda's interior structure, ocean dynamics, and habitability. As scientists continue to uncover the secrets of this hidden gem, Miranda stands as a reminder of the rich diversity of worlds waiting to be explored in the outer reaches of our solar system.

CHAPTER 5

The Role of Tidal Forces – Heating and Life on Uranus's Moons

Tidal forces are one of the most fascinating natural phenomena in planetary science. These forces, arising from gravitational interactions between celestial bodies, shape the surfaces, atmospheres, and even the internal structures of planets and moons. For Uranus's system, tidal forces play a critical role in driving geological activity and sustaining subsurface oceans on its moons. Among these moons, Miranda stands out as a prime example of how tidal forces can generate heat, potentially creating conditions suitable for life.

This chapter explores the mechanisms behind tidal forces, the specific processes heating Miranda's interior, and comparisons with other moons in the solar system that also rely on tidal forces to sustain their unique environments.

Tidal Interactions Explained: How Gravitational Forces Create Heat

Tidal forces arise from the gravitational pull exerted by a massive body—such as a planet—on a smaller orbiting object, like a moon. These forces are not uniform across the moon; the side facing the planet experiences a stronger gravitational pull than the far side. This difference creates a tidal bulge, stretching and deforming the moon.

1. The Basics of Tidal Interactions

The continuous stretching and squeezing of a moon due to tidal forces generate internal friction within its structure. This friction converts kinetic energy from the moon's orbital motion into thermal energy, heating its interior. The intensity of tidal heating depends on several factors, including:

- **The proximity of the moon to the planet**: Moons closer to the planet experience stronger gravitational forces, resulting in more significant tidal effects.

- **Orbital eccentricity**: A non-circular orbit causes the moon's distance from the planet to vary, intensifying the stretching and compressing as the gravitational pull fluctuates.

- **The moon's composition**: Moons with a higher proportion of deformable materials, such as water ice, are more susceptible to tidal heating.

2. The Role of Orbital Resonances

Tidal forces are often amplified by orbital resonances— gravitational interactions between multiple moons that lock them into specific orbital patterns. For example, when two moons are in a resonance, their gravitational pulls regularly reinforce each other, maintaining the eccentricity of their orbits. This sustained eccentricity ensures that tidal heating continues over long periods.

In the case of Uranus's system, the gravitational interactions between Miranda and other moons, such as Ariel and Umbriel, may create resonances that enhance tidal heating. However, due to the limited exploration of Uranus, the specific orbital dynamics in its system remain a topic of ongoing research.

3. Implications for Heat Generation

Tidal heating is a powerful mechanism that can significantly impact a moon's geology and habitability. On Earth, tidal forces drive the motion of ocean tides. On moons, however, the effects are far more dramatic, producing enough heat to sustain subsurface oceans or fuel volcanic activity. For Uranus's moons, tidal forces are a crucial factor in determining whether their interiors remain geologically active and whether liquid water can persist beneath their icy surfaces.

Miranda's Heating Process: How Tidal Forces Sustain a Subsurface Ocean

Miranda, the smallest of Uranus's major moons, offers a compelling case study in tidal heating. Despite its size, Miranda's surface is marked by dramatic geological features, including vast cliffs, ridged terrains, and coronae, all of which suggest a history of significant internal activity. Scientists believe that tidal forces are the primary driver of this activity, generating heat that may sustain a subsurface ocean beneath Miranda's icy crust.

1. Evidence of Tidal Heating

The unusual surface features of Miranda provide strong evidence for tidal heating. The presence of coronae—large, oval-shaped regions with ridges and valleys—suggests that material from the moon's interior has pushed upward, deforming the surface. These features are consistent with processes driven by internal heating, such as cry volcanism or tectonic activity.

Additionally, the chaotic terrain on Miranda's surface, characterized by overlapping regions of older and younger landscapes, indicates periods of resurfacing. This resurfacing could be linked to the movement of liquid water or slushy ice beneath the crust, facilitated by tidal heating.

2. Proximity to Uranus

Miranda's relatively close orbit around Uranus places it in a region where tidal forces are particularly strong. As the moon orbits, the gravitational pull from Uranus continuously stretches and compresses its interior. This process generates frictional heat, which may be sufficient to maintain a layer of liquid water beneath the icy crust.

3. The Role of Orbital Resonances

Miranda's past orbital interactions with other moons likely played a role in enhancing tidal heating. Scientists hypothesize that Miranda may have once been in a resonance with Ariel, another of Uranus's moons. This resonance would have increased the eccentricity of Miranda's orbit, amplifying tidal forces and generating more heat. Although Miranda's current orbit is nearly circular, suggesting that the resonance has ended, the heat generated during that period may have persisted, sustaining geological activity.

4. Subsurface Ocean Hypothesis

The combination of tidal heating and geological evidence supports the hypothesis that Miranda harbors a subsurface ocean. If such an ocean exists, it is likely located beneath an insulating ice shell. The thickness of this shell and the depth of the ocean depend on the balance between heat generation and heat loss. While direct evidence of an ocean has yet to be obtained, future missions to Uranus could confirm its existence through techniques such as ice-penetrating radar or magnetometer measurements.

Comparisons with Other Moons: Parallels with Europa and Enceladus

To better understand the role of tidal forces on Miranda, it is helpful to compare its heating processes and potential habitability with other moons in the solar system, particularly Europa and Enceladus. These moons, orbiting Jupiter and Saturn respectively, are among the most well-studied examples of tidal heating sustaining subsurface oceans.

1. Europa: A Model for Tidal Heating

Europa, one of Jupiter's largest moons, is widely regarded as one of the best candidates for extraterrestrial life in the solar system. Its subsurface ocean, hidden beneath an icy crust, is sustained by tidal heating resulting from its orbital resonance with two other Jovian moons, Io and Ganymede. This resonance maintains Europa's orbital eccentricity, ensuring continuous tidal flexing and heat generation.

The parallels between Europa and Miranda are striking. Both moons show evidence of resurfacing and geological activity linked to internal heating. However, Europa's larger size and more active orbital resonance provide a more robust mechanism for sustaining its ocean. Miranda's smaller size and weaker tidal forces mean that its ocean, if it exists, would likely be smaller and less stable.

2. Enceladus: A Cryovolcanic Wonder

Enceladus, a mid-sized moon of Saturn, has captured the attention of scientists due to its spectacular geysers, which spew water, ice, and organic molecules from its subsurface ocean into space. These geysers are powered by tidal heating, as Enceladus orbits Saturn in resonance with another moon, Dione. The discovery of these plumes has provided direct evidence of liquid water and potential habitability.

Miranda shares some similarities with Enceladus, particularly in its surface features and potential for cryovolcanism. While Miranda has not been observed to emit plumes, the ridged terrains and coronae on its surface suggest that internal processes driven by tidal heating could produce similar phenomena. If future missions were to detect plumes on Miranda, it would provide strong evidence of an active subsurface ocean.

3. Differences in Energy Sources

One key difference between Miranda and these other moons is the intensity of tidal heating. Europa and Enceladus benefit from strong orbital resonances and closer proximity to their host planets, resulting in higher levels of internal heat. Miranda, on the other hand, experiences weaker tidal forces, which may limit the size and longevity of its subsurface ocean. Despite this, the evidence of past activity on Miranda suggests that tidal heating has played a significant role in shaping its geological history.

4. Habitability Comparisons

In terms of habitability, Europa and Enceladus are considered more promising than Miranda due to their larger oceans and higher levels of energy availability. However, the potential for life on Miranda cannot be dismissed. If its subsurface ocean contains salts and other chemical nutrients, it could provide a stable environment for microbial life. The lower energy levels on Miranda might limit the complexity of potential life forms, but the basic ingredients for habitability could still exist.

Tidal forces are a powerful and transformative phenomenon in the solar system, capable of generating heat, driving geological activity, and sustaining subsurface oceans. For Uranus's moons, particularly Miranda, these forces offer a plausible explanation for the dynamic features observed on their surfaces and the potential for hidden oceans beneath their icy crusts.

Miranda, with its dramatic landscapes and evidence of internal heating, serves as a fascinating example of how tidal forces shape small, icy worlds. While its heating processes may be less intense than those of Europa or Enceladus, they are sufficient to raise the possibility of a subsurface ocean and even habitability. Future missions to Uranus could provide the data needed to confirm these hypotheses, shedding light on the role of tidal forces in creating environments where life might emerge.

As we continue to explore the outer solar system, Miranda reminds us of the incredible diversity of worlds and the powerful forces that shape them. Its story is one of resilience and transformation, offering a glimpse into the complex interplay of gravitational forces, geological processes, and the potential for life in the most unexpected places.

CHAPTER 6

Redefining Habitability in the Outer Solar System

When we think of habitability, our minds often drift to planets within the so-called "habitable zone" of their stars, where conditions are just right for liquid water to exist on the surface. However, discoveries in recent decades have shown that this conventional idea of habitability might be too narrow. In our solar system, moons orbiting the giant planets—particularly Jupiter, Saturn, and Uranus—demonstrate that habitability can arise in unexpected places, even far beyond the warmth of the Sun.

Uranus's major moons—Titania, Oberon, and Miranda—are captivating worlds in their own right. Each moon possesses unique characteristics, geological histories, and potentially even internal oceans that could offer the basic requirements for life. This chapter delves into what makes a moon habitable, examines the potential of Titania, Oberon, and Miranda to support life, and explores how these icy worlds challenge traditional notions of habitability.

What Makes a Moon Habitable? Criteria for Life-Supporting Environments

Defining habitability on icy moons requires us to think differently from how we traditionally evaluate planets like Earth. Instead of focusing solely on surface conditions, scientists look at subsurface oceans, internal heat sources, and the presence of essential chemical ingredients. These factors make certain moons prime candidates for life, despite their locations in the cold reaches of the outer solar system.

1. Liquid Water: The Foundation of Life

Liquid water is the cornerstone of habitability as we understand it. On Earth, wherever liquid water is found—whether in scorching hot springs, acidic lakes, or the frigid Antarctic ice shelf—life thrives. For icy moons, the prospect of liquid water exists not on the surface, where temperatures are too low, but beneath the icy crusts, where gravitational forces from the host planet generate heat through tidal flexing.

Tidal heating, created by the gravitational pull of a large planet on an orbiting moon, stretches and compresses the moon's interior, generating enough heat to melt ice and maintain liquid water beneath the crust. These internal oceans could provide stable environments where life might exist, shielded from harmful solar and cosmic radiation by the ice shell above.

2. Chemical Ingredients and Energy Sources

Life requires more than just water. Basic building blocks—carbon, hydrogen, nitrogen, oxygen, phosphorus, and Sulphur—must be present in accessible forms. These elements, often found in organic molecules, are necessary for creating the complex structures that make up living organisms. Additionally, life needs an energy source to power cellular functions. For icy moons, this energy could come from:

- **Chemical reactions between water and rock**: Subsurface oceans that interact with a rocky core could facilitate hydrothermal processes, similar to Earth's deep-sea vents, releasing nutrients and energy-rich compounds.

- **Radioactive decay**: The natural decay of radioactive elements within the moon's core can generate heat, contributing to a stable environment.

- **Tidal heating**: As previously mentioned, the flexing of a moon's interior due to gravitational interactions creates heat, which could drive chemical reactions beneficial for life.

3. Stability and Protection

A stable environment is crucial for sustaining life, and in the outer solar system, stability often depends on a thick ice shell that insulates a subsurface ocean from external conditions. This shell not only retains heat but also protects the ocean from solar and cosmic radiation. An additional protective factor could come from the planet's magnetosphere, which can shield moons from harmful radiation. For moons orbiting Uranus, the planet's magnetosphere could play an important role in maintaining conditions favorable for habitability.

Titania, Oberon, and Miranda's Potential: Exploring Each Moon's Unique Characteristics and How They Might Support Life

Uranus's three largest moons—Titania, Oberon, and Miranda—each have characteristics that make them intriguing candidates in the search for life. Although less studied than the moons of Jupiter and Saturn, they share similarities with Europa and Enceladus, which are known to have subsurface oceans. Let's examine each moon in detail.

1. Titania: The Largest of Uranus's Moons

As the largest moon of Uranus, Titania measures about 1,578 kilometers (980 miles) in diameter, making it slightly smaller than Jupiter's moon Europa. Titania's surface is a mixture of rocky and icy material, with a landscape characterized by canyons, craters, and fault lines.

Geological Activity and Potential for Liquid Water

Titania's surface shows signs of past geological activity, particularly in the form of fault lines and large canyons. The most notable of these is Messina Chasma, a massive canyon that stretches nearly 1,500 kilometers (930 miles) across the moon's surface. This tectonic activity suggests that Titania may have experienced internal heating in the past, potentially from tidal forces, which could have created conditions suitable for a subsurface ocean.

Although Titania's orbit is less eccentric than those of moons like Europa or Enceladus, it is still close enough to Uranus to experience some tidal heating, especially if it has engaged in orbital resonances with other moons in the past. Combined with the potential heat generated by radioactive decay within its core, tidal forces might sustain a thin layer of liquid water beneath Titania's icy crust.

Chemical Composition and Possibility for Life

Titania's surface composition includes water ice, carbon dioxide, and potentially ammonia, which acts as an antifreeze and could help maintain liquid water at lower temperatures. The presence of carbon dioxide suggests that Titania may have a source of carbon, a key element for life. If a subsurface ocean interacts with a rocky core, hydrothermal processes could release nutrients and support basic life forms.

2. Oberon: A World of Contrasts

Oberon, the second-largest moon of Uranus, is slightly smaller than Titania, with a diameter of approximately 1,523 kilometers (946 miles). Oberon is one of the darkest and most heavily cratered moons in the Uranian system, indicating an older, less geologically active surface. However, Oberon's size and composition suggest that it could still harbor an internal ocean.

Geological Features and Internal Heating

Unlike Titania, Oberon's surface shows fewer signs of tectonic activity, with a greater density of craters that suggests a geologically inactive history. However, Oberon has a few large fault valleys, indicating that some degree of internal activity may have occurred. If tidal heating or radioactive decay has persisted within Oberon's core, it could maintain a layer of liquid water beneath the surface.

Conditions for Life

The dark patches observed on Oberon's surface hint at organic compounds, possibly left by cometary impacts or generated from interactions within the moon itself. If Oberon has a subsurface ocean in contact with rock, chemical reactions between water and minerals could produce nutrients similar to those found around Earth's hydrothermal vents. These nutrient-rich environments could support microbial life, provided sufficient heat and stability exist.

3. Miranda: The Smallest and Most Dynamic

Despite being the smallest of Uranus's major moons, Miranda boasts some of the most unusual geological features in the solar system. Its patchwork surface includes towering cliffs, smooth plains, and three large oval-shaped regions called coronae, which are unique to Miranda.

Tidal Heating and the Potential for a Subsurface Ocean

Miranda's proximity to Uranus makes it particularly susceptible to tidal forces, which are believed to have driven its intense geological activity. Although Miranda's current orbit is nearly circular, scientists hypothesize that it may have been more eccentric in the past, allowing for stronger tidal interactions. This heating may have driven cry volcanism and tectonic processes, possibly creating a subsurface ocean beneath the ice.

Habitability and Potential for Life

The presence of coronae—regions of ridges and valleys—suggests that material from Miranda's interior may have risen to the surface. This process could indicate a reservoir of liquid or slushy ice beneath the crust, providing a stable environment shielded from radiation. If Miranda's Ocean contains salts or other conductive materials, it could generate an induced magnetic field in response to Uranus's magnetosphere, offering additional protection. Although direct evidence for life is lacking, the combination of internal heat, liquid water, and chemical interactions makes Miranda a compelling candidate for habitability.

Beyond the "Habitable Zone": How Uranus's Moons Challenge Traditional Ideas of Habitability

In traditional astronomy, the "habitable zone" refers to the region around a star where conditions are warm enough for liquid water to exist on a planet's surface. This concept, while useful, has been challenged by discoveries in the outer solar system, where icy moons orbiting gas giants exhibit environments capable of supporting life despite being far from the Sun. Uranus's moons offer an excellent case study in redefining habitability beyond the habitable zone.

1. The Role of Internal Heating

For moons outside the habitable zone, internal heating becomes the primary driver of habitability. Instead of relying on sunlight to maintain liquid water, these moons depend on tidal forces and radioactive decay to generate the necessary heat. This process is evident on Europa, Enceladus, and potentially Titania, Oberon, and Miranda. By reshaping our understanding of habitability, scientists now recognize that moons in the outer solar system could harbor life in subsurface oceans, completely isolated from sunlight.

2. Magnetospheric Protection in the Outer Solar System

In the outer reaches of the solar system, magnetospheric protection becomes crucial for preserving subsurface oceans and shielding potential life forms from harmful cosmic radiation. Uranus's magnetosphere extends far enough to encompass its major moons, providing a protective barrier against solar and cosmic radiation. This shielding effect, combined with the ice shell on each moon, creates a stable environment that could sustain liquid water and organic molecules over long timescales.

3. Implications for Exoplanetary Moons

The discovery of potentially habitable moons in the outer solar system has significant implications for the search for life beyond our solar system. Many exoplanets orbiting distant stars are gas giants, similar to Jupiter, Saturn, and Uranus. If these exoplanets have moon systems with similar tidal interactions and magnetospheric protection, they could harbor habitable environments, even if the planet itself lies beyond its star's habitable zone. This understanding broadens the scope of astrobiology, as researchers consider not only exoplanets but also their moons in the search for life.

4. Rethinking the Limits of Life

The presence of internal oceans on moons like Titania, Oberon, and Miranda challenges the limits of life as we know it. Life on Earth has proven remarkably adaptable, thriving in extreme environments such as deep-sea hydrothermal vents, acidic lakes, and polar ice caps. If life can exist in these harsh conditions, it may also survive in the icy oceans beneath Uranus's moons. These moons may lack sunlight, but they possess liquid water, energy sources, and essential nutrients—raising the tantalizing possibility that life could exist far from the traditional "habitable zone."

Titania, Oberon, and Miranda, the major moons of Uranus, each present a unique set of characteristics that make them intriguing candidates in the search for life. Their potential subsurface oceans, sustained by internal heating and shielded by thick ice shells, offer environments that could support microbial life. By studying these moons, scientists are redefining the concept of habitability, shifting the focus from surface conditions to hidden oceans and internal energy sources.

As we explore the outer solar system, moons like Titania, Oberon, and Miranda remind us that the search for life is not limited to Earth-like planets within the habitable zone. Instead, it extends to the icy worlds orbiting giant planets, where life may thrive in conditions vastly different from those on Earth. Future missions to Uranus and its moons could unlock the secrets of these hidden oceans, offering a glimpse into the potential for life in the most unexpected corners of our solar system and beyond.

CHAPTER 7

Rethinking Space Missions to Uranus

As the seventh planet from the Sun, Uranus is a giant of mystery in the outer solar system. Despite its unique characteristics, including its extreme axial tilt, faint ring system, and an array of fascinating moons, Uranus has been largely overlooked in planetary exploration. Since Voyager 2's flyby in 1986, no mission has returned to study this enigmatic world, leaving numerous questions unanswered about its atmosphere, magnetosphere, rings, and moons. Recent advancements in technology and a growing understanding of the potential habitability of icy moons have reignited scientific interest in Uranus, prompting calls for a dedicated mission.

This chapter explores the rationale behind a new mission to Uranus, proposed mission concepts for studying the planet and its moons, and the technological advancements that could enable a deeper exploration of this distant world.

The Case for a New Mission: Why Scientists Are Calling for Renewed Exploration of Uranus and Its Moons

Voyager 2's historic flyby provided humanity with its first glimpse of Uranus and its system of moons, rings, and magnetic field. However, the data collected during this brief encounter was limited, capturing only a snapshot of a complex world. Uranus is unique among the giant planets for its tilted rotation, unusual magnetic field, and a system of moons that may harbor subsurface oceans. For these reasons and more, scientists are eager to return to Uranus with modern instruments capable of addressing key questions about its formation, structure, and potential for supporting life.

1. Understanding Uranus's Atmosphere and Climate

Uranus has the coldest atmosphere in the solar system, with temperatures plunging as low as -371°F (-224°C). Its atmosphere, composed primarily of hydrogen, helium, and methane, presents unique challenges and mysteries. For example, Uranus emits far less heat than expected, especially compared to its neighbor, Neptune, which is nearly identical in size and composition. This lack of internal heat raises questions about the processes governing Uranus's atmospheric dynamics.

Voyager 2's data hinted at weather patterns and cloud bands within Uranus's atmosphere, but it lacked the resolution to provide a detailed picture. A dedicated mission could probe the depths of the atmosphere, studying cloud formation, circulation patterns, and seasonal changes. Additionally, understanding the chemistry of Uranus's atmosphere could offer insights into the planet's history, including the processes that led to its extreme tilt.

2. Investigating Uranus's Unique Magnetosphere

One of the most surprising discoveries made by Voyager 2 was Uranus's irregular magnetic field, which is tilted 59 degrees relative to the planet's rotational axis and offset from its center. This asymmetric magnetosphere behaves very differently from Earth's or Jupiter's, creating an environment of fluctuating particle radiation that interacts with Uranus's rings and moons. Understanding Uranus's magnetosphere is essential not only for planetary science but also for the study of magnetospheres in general, as it may offer clues to how magnetic fields operate on exoplanets with similarly tilted axes.

Additionally, Uranus's magnetosphere could play a critical role in maintaining the habitability of its moons, shielding them from harmful radiation and allowing stable environments to persist. A modern spacecraft equipped with advanced magnetometers could map the magnetosphere in detail, providing a clearer picture of how it interacts with the planet's atmosphere, rings, and moons.

3. Exploring the Potential Habitability of Uranus's Moons

Uranus's major moons—Titania, Oberon, and Miranda—are among the most intriguing objects in the outer solar system. They possess surfaces marked by canyons, ridges, and other features suggesting a history of tectonic and cryovolcanic activity. Recent research indicates that these moons may harbor subsurface oceans, kept warm by tidal heating. If true, this would make Uranus's moons prime candidates for Astro biological studies, as subsurface oceans are considered potential habitats for microbial life.

The discovery of potentially habitable environments on these moons has elevated their scientific significance, positioning Uranus's moons alongside Europa, Enceladus, and Titan as key targets for the search for life beyond Earth. A mission to Uranus would allow scientists to assess the moons' geological activity, surface composition, and potential for liquid water beneath the ice.

4. Completing Our Understanding of the Ice Giants

Uranus and Neptune are often referred to as the "ice giants" due to their high proportions of water, ammonia, and methane, distinguishing them from the "gas giants" Jupiter and Saturn. However, compared to the gas giants, the ice giants are poorly understood. NASA's focus on Mars, Jupiter, and Saturn has left a knowledge gap regarding the ice giants, limiting our understanding of planetary formation and evolution across the solar system.

A dedicated mission to Uranus would provide valuable insights into the differences between gas giants and ice giants, helping scientists refine their models of planetary formation. Understanding these processes is crucial not only for studying our solar system but also for interpreting the diversity of exoplanets observed around other stars.

Proposed Mission Concepts: Ideas for Future Missions, Including Orbiters, Landers, and Probes

Given the scientific potential of a mission to Uranus, space agencies have proposed various mission concepts designed to explore the planet and its moons in detail. Each concept comes with its unique advantages, challenges, and scientific objectives. Here are some of the leading ideas for future Uranus missions.

1. Uranus Orbiter with Atmospheric Probes

An orbiter mission to Uranus, equipped with multiple atmospheric probes, is one of the most widely discussed concepts. Such a mission could spend years orbiting Uranus, conducting comprehensive studies of the planet, its rings, magnetosphere, and moons. Key components of this mission concept include:

- **Atmospheric probes**: Dropped into Uranus's atmosphere, these probes could measure pressure, temperature, chemical composition, and wind speeds at various depths. Probes could provide a direct sample of the atmosphere, offering insights that remote sensing alone cannot achieve.

- **High-resolution cameras and spectrometers**: An orbiter equipped with high-resolution imaging and spectroscopic instruments could map Uranus's surface features, study cloud layers, and analyze atmospheric composition.

- **Magnetometer and particle instruments**: These instruments could map Uranus's magnetic field and measure particle radiation in the magnetosphere, providing valuable data on its structure and interactions with the planet's rings and moons.

76

2. Titania and Oberon Lander Missions

While orbiters offer a broad view of Uranus's system, landers could provide in-depth analysis of specific moons. Titania and Oberon, the two largest moons, are particularly appealing candidates for lander missions due to their potential for subsurface oceans and geological features that suggest past tectonic activity. Key objectives of a lander mission could include:

- **Geological sampling**: A lander could analyze surface materials, providing clues about the moon's geological history and internal composition.

- **Cryobot exploration**: A cryobot, or ice-penetrating probe, could potentially drill through the icy crust, reaching subsurface layers. If an ocean exists beneath the ice, the cryobot could collect samples, testing for signs of life.

- **Seismometers**: These instruments could detect seismic activity, helping scientists understand the internal structure and ongoing geological processes of the moons.

3. Flyby and Reconnaissance Missions

For missions with limited budgets or timelines, a flyby mission offers a more feasible approach to studying Uranus and its moons. Similar to Voyager 2, a flyby mission could collect high-quality images and data during a single pass. Advances in technology mean that a modern flyby could gather far more information than Voyager 2, including better imaging, spectrometry, and magnetic data.

4. Miranda Subsurface Ocean Explorer

Miranda, the smallest of Uranus's major moons, presents a unique scientific opportunity due to its distinctive surface features, which suggest a history of intense geological activity. A dedicated mission to Miranda could include a lander or ice-penetrating probe designed to investigate the possibility of a subsurface ocean. This mission could involve:

- **Surface analysis**: A lander equipped with spectrometers and other analytical tools could study the composition of Miranda's surface, identifying potential cryovolcanic materials and other indicators of internal activity.

- **Ice-penetrating radar**: This tool would allow scientists to search for subsurface structures, including any liquid water reservoirs that might exist beneath the surface.

- **Cryobot deployment**: Similar to concepts proposed for Europa and Enceladus, a cryobot could attempt to penetrate Miranda's crust, reaching potential liquid layers that may harbor microbial life.

5. Uranus System Sample Return Mission

A sample return mission from Uranus's system, while challenging, could yield unparalleled scientific insights. By collecting materials from Uranus's atmosphere, rings, or moons, such a mission would provide researchers on Earth with tangible samples to study in detail. Although technically ambitious, sample return missions are becoming more feasible with advancements in autonomous spacecraft and robotic technology.

Technological Advances: How Today's Technology Could Reveal New Insights into Uranus's Environment and Moons

Technological innovations over the past few decades have revolutionized planetary exploration, enabling scientists to study distant worlds with unprecedented precision. For a mission to Uranus, these advancements would allow a much deeper exploration than was possible during Voyager 2's flyby. Here's a look at some of the technologies that could make a Uranus mission successful.

1. Autonomous Navigation and AI-Driven Data Processing

Given the vast distance to Uranus, a spacecraft traveling there would require sophisticated onboard systems capable of autonomous navigation and data processing. Unlike closer targets such as Mars, communication delays with Uranus are significant, ranging from 2.5 to 5 hours each way. Autonomous systems powered by artificial intelligence (AI) could enable the spacecraft to make decisions in real-time, adapting to unexpected conditions and ensuring mission success.

2. Advanced Power Sources

One of the major challenges of a Uranus mission is the lack of sunlight at such a distant location. Unlike missions to Mars, which can rely on solar panels, a Uranus mission would require an alternative power source, such as a radioisotope thermoelectric generator (RTG) or advanced nuclear power sources. NASA has been developing newer, more efficient RTGs that could provide sufficient power for an orbiter, lander, or probe.

3. Ice-Penetrating Technology

One of the most exciting prospects of a Uranus mission would be the exploration of its moons' subsurface oceans. Advanced ice-penetrating radar, cryobots, and other subsurface exploration tools could enable a spacecraft to detect and analyze liquid water reservoirs beneath the moons' icy crusts. These technologies, initially developed for Europa and Enceladus missions, could be adapted for the unique conditions of Uranus's moons.

4. High-Resolution Imaging and Spectrometry

Modern imaging and spectrometric instruments would provide far more detailed information about Uranus's atmosphere, rings, and moons. High-resolution cameras, infrared and ultraviolet spectrometers, and other advanced sensors could capture images of Uranus's clouds, detect compounds in the atmosphere, and analyze surface compositions in unprecedented detail. These instruments would allow scientists to build a comprehensive picture of Uranus's chemical and geological makeup.

5. Advanced Magnetometers and Particle Detectors

Studying Uranus's unique magnetosphere would require sophisticated magnetometers and particle detectors capable of mapping the magnetic field and analyzing its interactions with solar wind and the planet's moons. Modern magnetometers can detect subtle variations in magnetic fields, allowing scientists to model the magnetosphere in much greater detail than was possible with Voyager 2.

The case for a dedicated mission to Uranus has never been stronger. With its unusual magnetosphere, intriguing atmosphere, and potentially habitable moons, Uranus represents a critical gap in our understanding of the outer solar system. Proposed mission concepts—including orbiters, landers, and probes—offer various pathways to explore this ice giant and its moons, each with unique scientific goals and technical challenges.

Technological advances in autonomous navigation, power systems, ice-penetrating technology, high-resolution imaging, and magnetometry have opened new possibilities for studying Uranus in detail. A mission to Uranus could not only reveal the secrets of this often-overlooked planet but also expand our knowledge of planetary formation, atmospheric dynamics, and the potential for life in icy environments.

As space agencies like NASA, ESA, and others consider future missions to the outer solar system, Uranus stands as a prime candidate for exploration. A dedicated mission to this icy world would complete our exploration of the gas and ice giants, filling critical gaps in our knowledge of the solar system and offering new insights into the nature of distant, potentially habitable moons.

CHAPTER 8

Uranus and the Evolution of Planetary Science

Uranus has long been a planet of intrigue and mystery within our solar system. As an ice giant with unique features like an extreme axial tilt, an offset magnetic field, and a family of moons that may harbor subsurface oceans, Uranus stands out among the planets. However, it has been one of the least explored worlds in our cosmic neighborhood, largely because it lies at the edge of practical reach for interplanetary missions. The 1986 flyby of Voyager 2 provided a brief but invaluable glimpse into this distant world, capturing data that has since reshaped our understanding of planetary science. However, recent advances in technology and data analysis have allowed scientists to revisit Voyager 2's findings, revealing new insights that continue to fuel the drive for further exploration.

This chapter delves into the transformative impact of the Voyager 2 mission to Uranus, the importance of data reanalysis in modern planetary science, and the lessons Uranus has taught us about the evolving field of space exploration. As technology continues to advance, the prospects for future missions to Uranus become increasingly promising, holding the potential to answer some of the most profound questions about our solar system and beyond.

Lessons from Uranus: How Revisiting Voyager 2 Data Has Transformed Our Understanding

When Voyager 2 completed its flyby of Uranus on January 24, 1986, it marked the first—and, to date, only—time humanity has visited the ice giant. Voyager 2's brief encounter with Uranus lasted mere hours, yet the data it collected transformed our understanding of this enigmatic planet and its moons. Before Voyager 2, Uranus was thought to be relatively uninteresting, with a bland, featureless appearance. However, Voyager revealed a complex and dynamic system that has continued to captivate scientists for decades.

1. Uncovering Uranus's Unique Atmosphere

Voyager 2's observations revealed that Uranus has the coldest atmosphere in the solar system, with temperatures dropping to around -371°F (-224°C). Composed primarily of hydrogen, helium, and methane, this atmosphere is marked by a blue-green hue due to methane's absorption of red light. However, what puzzled scientists was the lack of internal heat emissions from Uranus, especially when compared to its neighboring ice giant, Neptune. Unlike Neptune and other gas giants, Uranus emits almost no excess heat from its core, leading scientists to question why this planet appears so thermally "inactive."

Voyager's data prompted scientists to explore theories on why Uranus lacks internal heat. Some hypotheses suggest that an early impact may have disrupted the internal heat transfer mechanisms within Uranus, while others propose that a unique composition or structure within its atmosphere and core limits heat escape. These theories have not only deepened our curiosity about Uranus but have also expanded our understanding of planetary formation and thermal dynamics, challenging scientists to rethink their models of gas and ice giants.

2. The Discovery of a Tilted and Asymmetric Magnetic Field

One of Voyager 2's most surprising discoveries was Uranus's magnetic field, which is both tilted at an angle of 59 degrees to the planet's rotational axis and offset from its center. Unlike Earth, Jupiter, and Saturn, where the magnetic field aligns more closely with the rotational axis, Uranus's magnetic field behaves unpredictably, forming a complex and asymmetrical magnetosphere. This irregularity creates an environment with unique radiation dynamics, influencing Uranus's atmosphere, rings, and moons.

The discovery of Uranus's unusual magnetic field reshaped our understanding of planetary magnetism. It demonstrated that magnetic fields could be generated by mechanisms different from those found on Earth, where the molten iron core produces a relatively stable and aligned field. Scientists now believe that Uranus's magnetic field may originate from a "slush layer" of water, ammonia, and methane that is highly conductive and lies beneath its icy surface. This idea has inspired researchers to consider that other planets and exoplanets may also have magnetospheres that operate in unconventional ways.

3. Uncovering the Mysteries of Uranus's Moons

Voyager 2 also provided humanity's first close-up views of Uranus's moons, including Titania, Oberon, and Miranda. These moons, thought to be simple icy worlds, revealed unexpected geological features. Miranda, in particular, displayed a chaotic landscape with towering cliffs, ridged plains, and large oval structures known as coronae, suggesting a history of intense geological activity. This revelation prompted scientists to consider the role of tidal heating in maintaining subsurface oceans on these moons, much like the processes seen on Jupiter's Europa and Saturn's Enceladus.

The discovery of possible geological activity and tectonic processes on Uranus's moons has led scientists to reconsider the potential for habitability in the outer solar system. If tidal heating and internal friction can sustain subsurface oceans beneath icy crusts, these moons could potentially support life, even at the cold outer edges of our solar system. Voyager 2's images and data have therefore reshaped the criteria for habitability, suggesting that liquid water and energy sources might exist on moons far from the Sun.

The Importance of Data Reanalysis: How New Insights Can Emerge from Older Data with Modern Technology

The Voyager 2 mission produced a wealth of data, but the technology available in the 1980s limited scientists' ability to analyze it fully. Over the decades, advancements in computational power, data processing algorithms, and imaging technology have enabled scientists to revisit Voyager's data, uncovering insights that were previously inaccessible.

1. Enhanced Imaging and Data Resolution

One of the most significant advancements has been in image processing technology. High-resolution reanalysis of Voyager 2's images has allowed scientists to detect finer details within Uranus's atmosphere, rings, and moons. For example, enhanced imaging has revealed subtle cloud bands and potential storm systems within Uranus's atmosphere, hinting at previously undetected atmospheric dynamics. These findings challenge the initial perception of Uranus as an inert planet and suggest that, like Jupiter and Saturn, it may experience seasonal and weather-related changes over time.

Advanced imaging has also provided more detailed views of the moons, revealing previously unseen features on Titania, Oberon, and Miranda. This technology has allowed scientists to refine their geological interpretations of these moons, identifying possible cryovolcanic flows and tectonic fractures. The refined images have furthered our understanding of these moons' geological histories and the potential for subsurface oceans.

2. Improved Magnetospheric Analysis

New developments in data processing have enabled scientists to model Uranus's magnetosphere in unprecedented detail. By reanalyzing Voyager 2's magnetometer data, researchers have gained insights into the structure of Uranus's magnetic field, including its strength, shape, and variations over time. Modern simulations have helped scientists understand how Uranus's magnetosphere interacts with the solar wind, its moons, and the planet's own rotation.

This reanalysis has had broader implications for understanding magnetospheres throughout the solar system and beyond. Scientists are now using these insights to model the magnetic fields of exoplanets, particularly those with unusual tilts or orbits. By learning more about Uranus's magnetosphere, researchers are better equipped to interpret the behavior of magnetic fields on distant worlds, expanding our knowledge of planetary physics.

3. New Atmospheric and Chemical Models

Voyager 2's data included infrared and ultraviolet spectrometry, which allowed scientists to analyze Uranus's atmospheric composition. With modern analytical tools, scientists have been able to develop more accurate models of the planet's atmosphere, including its layered structure, composition, and potential chemical reactions occurring within it. These models suggest that Uranus's atmosphere may have a distinct composition from that of Jupiter, Saturn, or even Neptune, which could affect its thermal properties and the formation of clouds and storms.

Modern chemical models have also revealed insights into Uranus's apparent lack of internal heat. Theories regarding the suppression of internal heat, such as a barrier layer within the atmosphere or a disruption in the planet's core, are being explored through advanced modeling. These findings not only advance our understanding of Uranus but also influence our understanding of other icy planets and exoplanets.

4. Revisiting the Potential for Habitability on Uranus's Moons

Recent advancements in data processing have allowed scientists to revisit the geology and composition of Uranus's moons, enhancing our understanding of their potential habitability. By reanalyzing spectral data, researchers have been able to detect signs of water ice, organic compounds, and possible salts on these moons. These compounds are crucial for sustaining life and could serve as indicators of subsurface oceans beneath the moons' icy crusts.

In particular, modern data analysis techniques have allowed scientists to model the thermal and gravitational interactions between Uranus and its moons. These models suggest that tidal heating, coupled with potential chemical reactions between water and rock, could create stable environments beneath the icy surfaces of Titania, Oberon, and Miranda. This finding has expanded the search for life, indicating that even distant, frozen moons might support habitable environments if they possess liquid water, organic compounds, and energy sources.

The Future of Exploration: What Uranus and Its Moons Can Teach Us About the Evolving Field of Planetary Science

The discoveries and insights derived from Voyager 2's data have not only deepened our understanding of Uranus but also reshaped the field of planetary science. As we look toward future missions, Uranus and its moons offer valuable lessons that will shape the direction of space exploration for decades to come.

1. Expanding the Concept of the Habitable Zone

Traditionally, the "habitable zone" around a star has been defined as the region where liquid water can exist on a planet's surface, assuming Earth-like conditions. However, findings from Voyager 2 and subsequent reanalysis have shown that the habitable zone may be more flexible than initially thought. Icy moons like those orbiting Uranus could harbor liquid water beneath their surfaces, thanks to tidal heating and internal chemical processes. This expanded concept of habitability has profound implications for the search for life beyond Earth, suggesting that even cold, distant worlds could be home to microbial ecosystems.

2. Pioneering Techniques in Data Reanalysis

The reanalysis of Voyager 2 data has demonstrated the value of archival data in planetary science. As new technologies emerge, older data sets can yield fresh insights, often at a fraction of the cost of launching a new mission. This approach has inspired a movement within the scientific community to reexamine data from past missions to other planets, including Mars, Jupiter, and Saturn. Data reanalysis has become a crucial tool for planetary science, maximizing the scientific return on investment and continually enhancing our understanding of the solar system.

3. Innovations in Mission Design and Technology

The need to explore Uranus has spurred advancements in mission design and technology, leading to the development of new instruments, power systems, and spacecraft architectures capable of withstanding the extreme conditions of the outer solar system. Technologies such as ice-penetrating radar, advanced magnetometers, and autonomous navigation systems are now being designed for future Uranus missions, including proposed orbiters and landers.

These innovations will not only benefit missions to Uranus but also enable the exploration of other distant worlds, such as Neptune and Kuiper Belt objects. The lessons learned from designing missions to Uranus are shaping the next generation of space exploration, expanding the reach of humanity's robotic explorers to the most remote and mysterious parts of our solar system.

4. Preparing for Exoplanetary Exploration

Uranus's unique magnetic field, tilted rotation, and icy composition make it a valuable analog for studying exoplanets, particularly those known as "super-Earths" and "mini-Neptunes" that orbit other stars. By studying Uranus, scientists can refine their understanding of planetary formation, atmospheric dynamics, and magnetic fields in icy planets. These insights are crucial for interpreting observations of exoplanets, helping astronomers determine which of these distant worlds might be habitable.

As we continue to discover exoplanets with extreme tilts, eccentric orbits, and magnetic anomalies, Uranus serves as a local laboratory where scientists can test theories and develop techniques applicable to planets beyond our solar system.

The Voyager 2 mission to Uranus, and the subsequent decades of data reanalysis, have transformed Uranus from a distant enigma into a fascinating world of scientific intrigue. Through lessons learned from Uranus's atmosphere, magnetic field, and icy moons, planetary science has evolved, embracing the complexity and diversity of worlds beyond the traditional "habitable zone."

With advances in technology and a renewed focus on outer solar system exploration, the next steps in Uranus research hold promise for uncovering some of the deepest mysteries of our solar system. Future missions to Uranus could revolutionize our understanding of ice giants, offer insights into planetary formation, and potentially reveal the conditions needed for life in unexpected places. As we stand on the cusp of a new era of exploration, Uranus and its moons remind us that the quest for knowledge extends far beyond the familiar, pushing us to explore and redefine the boundaries of our cosmic neighborhood.

CONCLUSION

Magnetic Unknown Handbook Of Uranus

As we reach the conclusion of *The Magnetic Unknown Handbook of Uranus*, we find ourselves standing on the brink of a vast, uncharted realm in planetary science. Uranus, the seventh planet from the Sun, has captivated scientists and dreamers alike, holding tightly to its secrets within its icy depths, tilted magnetic field, and enigmatic moons. Despite being one of the least explored planets in the solar system, Uranus presents mysteries that challenge our understanding and redefine the possibilities of the universe we inhabit.

The journey through this book has taken us across countless scientific revelations, from the unique characteristics of Uranus's atmosphere to the surprising behaviors of its magnetosphere, and the compelling potential of its moons to harbor subsurface oceans. Each chapter has unfolded a new layer of complexity in this ice giant, highlighting how much more there is to learn. This conclusion serves not just as a summary of those revelations but also as a call to action, urging us to continue our pursuit of knowledge about Uranus and its place in our solar system.

The Case for Uranus Exploration

If there's one overriding theme that emerges from the study of Uranus, it is the need for a dedicated mission to this distant planet. The Voyager 2 flyby in 1986 provided a fleeting, incomplete glimpse of Uranus and its moons, sparking more questions than it answered. That single encounter, though brief, altered our understanding of the outer solar system by revealing Uranus as an active, dynamic world with a magnetic field unlike any other and moons that may contain the conditions necessary for life.

In recent decades, technology has advanced tremendously, enabling us to revisit the data collected by Voyager 2 with new analytical tools. Enhanced image processing, improved spectrometric analysis, and advanced magnetospheric modeling have helped scientists make strides in understanding Uranus, but these techniques alone cannot replace the need for direct exploration. There is an urgency to return to Uranus, especially in light of recent discoveries about icy moons and subsurface oceans in other parts of the solar system. A mission to Uranus could redefine our concept of habitability and uncover conditions that might support microbial life in environments previously deemed inhospitable.

Uranus's Lessons for Planetary Science

Uranus has served as a teacher, challenging planetary scientists to rethink their assumptions about planets and moons, magnetic fields, and the criteria for habitability. Through studying Uranus's offset and tilted magnetic field, researchers have developed new models for understanding how magnetic fields might behave on other planets. This knowledge is critical as we expand our search for exoplanets and seek to understand the magnetic environments that may shield them from cosmic radiation, allowing life to exist on their surfaces or beneath icy crusts.

The extreme axial tilt of Uranus, which causes its poles to face the Sun at certain points in its orbit, has redefined our understanding of seasonal dynamics in planetary atmospheres. By examining Uranus's unusual rotation, scientists have begun to consider the effects of axial tilt on climate, atmospheric circulation, and magnetic field orientation, not only in our solar system but also in distant star systems. This knowledge is crucial as we study exoplanets with similar tilts and strive to predict their climate cycles, atmospheric dynamics, and potential to support life.

Perhaps one of the most profound insights Uranus offers is the concept of habitability beyond the traditional "habitable zone." The moons of Uranus, particularly Titania, Oberon, and Miranda, have shown that liquid water and energy sources can exist far from the Sun, under thick ice shells that provide protection from radiation. The discovery of subsurface oceans on these icy moons has expanded our understanding of where life might exist, leading scientists to consider that life could be sustained by internal heat rather than sunlight. This paradigm shift has spurred researchers to search for life in places previously considered unlikely, from the icy moons of Jupiter and Saturn to distant exoplanets orbiting faint, cool stars.

The Power of Data Reanalysis and Technological Innovation

As we've explored in this book, the power of data reanalysis has been essential in expanding our knowledge of Uranus. By revisiting Voyager 2's original data with modern technology, scientists have managed to refine our understanding of Uranus's atmosphere, magnetic field, and moon system. These advancements remind us that planetary science is an evolving field where even data collected decades ago can reveal new insights with the right tools.

Reanalysis has breathed new life into the Voyager 2 data, showing that scientific progress is often a combination of patience, curiosity, and technological advancement.

The potential for new discoveries on Uranus serves as a reminder of the importance of continuous innovation in space exploration. Recent advancements in autonomous navigation, artificial intelligence, and robotic technology have made missions to the outer solar system more feasible and more informative than ever before. High-powered magnetometers, ice-penetrating radars, spectrometers, and advanced imaging systems could provide scientists with unprecedented views of Uranus's atmosphere, rings, and moons, transforming our understanding of this distant world. The next generation of missions will rely on this new technology to explore uncharted regions of space, continuing the legacy of Voyager 2.

Expanding the Frontiers of Human Knowledge

Uranus remains one of the most mysterious worlds in our solar system, but it is also one of the most promising. A return to Uranus could allow scientists to fill gaps in our understanding of planetary formation and evolution, reveal the secrets of the ice giants, and even explore the possibility of life beyond Earth. Each moon, with its unique geological features and potential subsurface oceans, offers a new opportunity to explore questions that are central to humanity's quest for knowledge: How did the planets and moons form? What role does magnetism play in shaping planetary environments? Could life exist in forms and places we have yet to imagine?

A mission to Uranus would contribute to our broader understanding of the solar system, providing insights into the early stages of planetary development and the forces that shape planets and moons over billions of years. This knowledge has implications beyond Uranus, helping us interpret the structure and behavior of exoplanets and potentially laying the groundwork for understanding habitable environments elsewhere in the universe. In this sense, Uranus is a gateway to a larger understanding of the cosmos, bridging the familiar worlds closer to the Sun with the mysterious, icy giants at the edge of our solar system.

The Call to Future Exploration

The journey to uncover the mysteries of Uranus is far from over. As we continue to learn from this unique planet and its system of moons, the need for a new mission becomes ever more compelling. A dedicated mission to Uranus, equipped with state-of-the-art instruments, could provide answers to questions that have puzzled scientists for decades, revealing details about the planet's interior, the structure of its magnetic field, and the potential habitability of its moons. This mission would not only expand our understanding of Uranus but also inspire future generations of explorers, scientists, and engineers to push the boundaries of what is possible.

In closing, *The Magnetic Unknown Handbook of Uranus* has aimed to shed light on one of the least understood planets in our solar system. Uranus challenges our expectations, redefines our concepts of habitability, and demonstrates the power of data, technology, and human curiosity to unlock the secrets of distant worlds. As humanity continues to reach for the stars, Uranus stands as a reminder of the vast unknowns that still await us and the thrilling potential that comes with exploring the uncharted corners of our solar system.

The quest for knowledge is an ever-evolving journey, and Uranus reminds us that some of the most profound discoveries may lie in the most unexpected places. Whether in the magnetic fields that defy alignment, the moons that may conceal oceans, or the icy surface that belies dynamic processes within, Uranus invites us to look deeper, ask bigger questions, and embrace the unknown. For as long as we remain curious, the universe will always have mysteries to offer, with Uranus standing as a testament to the wonders that await those willing to seek them out.